NUREG-0484
Rev. 1

I0493923

Methodology for Combining Dynamic Responses

R. K. Mattu

Office of
Nuclear Reactor Regulation

U.S. Nuclear Regulatory
Commission

NUREG-0484
Rev. 1

Methodology for Combining Dynamic Responses

Manuscript Completed: April 1980
Date Published: May 1980

R. K. Mattu

Division of Safety Technology
Office of Nuclear Reactor Regulation
U.S. Nuclear Regulatory Commission
Washington, D.C. 20555

ABSTRACT

The NRC has historically required that the structural/mechanical responses due to various accident loads and loads caused by natural phenomena (such as earthquakes) be combined when analyzing structures, systems, and components important to safety. This requirement flows from 10CFR50, Appendix A, General Design Criterion 2, which calls for an appropriate combination of the effects of the above events to be reflected in the design bases of safety equipment. This requirement has been implemented in various ways both within the NRC and the Nuclear Industry. An NRR Working Group constituted to examine load combination methodologies developed recommendations which were published in September 1978 as NUREG 0484. NRC adopted the Working Group's recommendations which stated: "For the combination of the dynamic responses within the Reactor Coolant Pressure Boundary (RCPB) and its supports which result from the coincidence of an SSE and LOCA, the Square Root of the Sums of the Squares (SRSS) technique is acceptable contingent upon performance of a linear elastic dynamic analysis to meet the appropriate ASME Code, Section III, Service Limit." Revision 1 of NUREG-0484 extends the conclusions of the original NUREG-0484 on the use of SRSS methodology for the combination of SSE and LOCA responses beyond RCPB to any other ASME Section III, Class 1, 2, or 3 affected system, component or support, and provides criteria for the combination of dynamic responses other than SSE and LOCA.

CONTENTS

1. INTRODUCTION

(A) Objective of Report

NUREG-0484 Revision 1, Methodology for Combining Dynamic Responses supersedes NRC Staff Working Group Report NUREG-0484 published in September 1978. This revision does not alter the conclusions of NUREG-0484. It extends the conclusions of NUREG-0484 on the use of SRSS methodology for the combination of SSE and LOCA responses beyond the Reactor Coolant Pressure Boundary to any other ASME Section III, Class 1, 2, or 3 affected system, component, or support, and provides criteria for the combination of dynamic responses other than SSE and LOCA.

(B) Background

The NRC has historically required that the structural/mechanical responses due to various accident loads and loads caused by natural phenomena (such as earthquakes) be combined when analyzing structures, systems, and components important to safety. This requirement flows from 10 CFR Part 50, Appendix A, General Design Criterion 2, which calls for an appropriate combination of the effects of the above events to be reflected in the design bases of safety equipment. The requirement has been implemented in various ways both within the NRC and the Nuclear Industry.

Loads due to postulated accidents and natural phenomena often yield dynamic responses of short duration and rapidly varying amplitude in the structures and components exposed to the loads. For some loading phenomena, the accident analysis provides a definitive time history response and allows a straight-forward addition of responses where more than one load is acting concurrently. In other cases, no specified time-phased relationship exists, either because the loads are random in nature (e.g., vent chugging in pressure-suppression containments) or because the loads have simply been postulated to occur together (e.g., LOCA and SSE) without a known or defined coupling. Where a defined time-phase relationship is lacking, system designers have utilized several approaches to account for the potential interaction of the loads. One approach, the so-called absolute or linear summation (ABS) method, linearly adds the absolute values of the peak structural responses due to the individual dynamic loads. Babcock and Wilcox (B&W) and Combustion Engineering (CE) use the ABS method to analyze the primary coolant system components and supports. In general, the ABS method has also reflected the staff's conservative preference for the combination of dynamic load responses.

A second approach, referred to as SRSS, yields a combined response equal to the square root of the sum of the squares of the peak responses due to the individual dynamic loads. Depending on the relative magnitudes of the loads, the SRSS value will vary between about 0.7 ABS and 1.0 ABS for the combination of two dynamic loads. The SRSS method has been used by both General Electric (GE) and Westinghouse in selected applications. The NRC staff has approved the use of SRSS to combine certain modal responses caused by earthquakes[1] and

---

[1]Regulatory Guide 1.92, Combining Modal Responses and Spatial Components in Seismic Response Analysis, February 1976.

to combine LOCA and SSE loads in the analysis of PWR fuel bundles.[2] The staff also has been requested to consider its application in Mark I and Mark II containment pool dynamic loads evaluations.

The lack of a physical relationship between some of the loads has raised questions as to the proper methodology to be used in the design of nuclear power plants. The ABS approach may lead to overly conservative design requirements and may result in a system more rigid than that required to provide an optimum response to the thermal stresses present in normal operation. These overly conservative design requirements may also require unnecessary modifications for operating plants (if reanalyzed) which were licensed before promulgation of the present GDC 2. On the other hand, there needs to be a technical basis for the use of the SRSS method.

An NRR Working Group was constituted in June 1978 to examine load combination methodologies and to develop a recommendation concerning criteria or conditions for their application. During the course of its deliberations, the Working Group met with each of the NSSS vendors, and telephone calls were made to solicit the views of a number of architect/engineering firms. A limited amount of time was made available to this group so that the work reported in NUREG-0484, published in September 1978, focused its attention on application of the methodologies to the primary reactor coolant system. In this regard, another consideration was the amount of information available at that time regarding the design and response of various primary system components to

----

[2]Memo, R. Meyer to R. Mattson, Use of SRSS Load Combination for Fuel, dated July 18, 1978.

dynamic loads. Emphasis was placed on an objective review of each methodology and the development of a balanced judgment with regard to the information available at that time. The Working Group efforts focused on the ABS and SRSS methods and their evaluations of these methods are repeated in Sections 2(A) and 2(B) of this report. No other combination methodology was identified by the Working Group that warranted further investigation at that time. Although not a specific task of the group, decoupling of LOCA and SSE loads was considered an inseparable part of the discussions and is discussed in Section 2(C).

(C) Report Content

Section 2 of this report summarizes the work done by the Working Group in 1978 for the combination of the dynamic responses within the Reactor Coolant Pressure Boundary (RCPB) and its supports, which result from the coincidence of an SSE and LOCA. Section 3 provides methodology and its rationale for combining dynamic responses due to SSE and LOCA to any ASME Class 1, 2, and 3 affected system, component, or support. Section 4 describes the criteria for combination of dynamic responses other than those of SSE and LOCA. The staff conclusions are provided in Section 5. Sections 3, 4, and 5 reflect the staff work and findings completed since the publication of NUREG-0484 in September 1978. Section 6 describes briefly continuation of generic effort on load/ response combinations.

2. <u>DYNAMIC RESPONSES AND COMBINATION METHODOLOGIES</u> - (NUREG-0484)

ABS and SRSS methods and their evaluation are discussed below. Also discussed here are decoupling of LOCA and SSE loads/responses.

(A) <u>Absolute or Linear Summation Method</u>

<u>Concept</u>

In the design of nuclear power plant components when the time-phase relationship between the responses caused by two sources of dynamic loading is undefined or random, the use of the ABS methodology linearly adds the peaks of the individual colinear responses to obtain the combined response. No consideration is given to the probability of occurrence of either of the associated events that cause the combined loading in arriving at the procedure for combining the randomly phased responses. An example of a design equation using the ABS method for the combination of SSE and LOCA loads is provided below (N represents normal loads):

$$\left|N\right| + \left|SSE\right| + \left|LOCA\right| \leq \text{Level D Service Limit}$$

<u>Advantages/Disadvantages</u>

(1) The ABS method generally provides greater margin to failure as compared to other methods considered, although it may be overly conservative given the random dynamic nature of the responses. The extra margin to failure

obtainable by the ABS method in relation to the SRSS method is a function of the relative magnitude of the dynamic responses. The maximum difference between ABS and SRSS for combining two responses is approximately 30 percent and occurs when the peak amplitudes of the dynamic responses which are being added are equal.

(2) The ABS method is conceptually straightforward in that it attempts to bound the combined response by adding the individual peak responses. Application or justification of the method does not involve probabilistic bases.

(3) The ABS method is acceptable to the NRC staff, although consideration has been given to the use of SRSS on a case-by-case basis. Where SRSS has been used in the design of existing plants or where new loading conditions have been identified, application of ABS could require extensive and sophisticated reanalyses or plant modifications in order that response criteria be met.

(4) The use of the ABS method may result in more rigid systems which is not beneficial when the design must consider thermal stresses.

(B) <u>Square Root Sum of Squares Method (SRSS)</u>

Concept

The SRSS approach of combining dynamic transients is to square the peak value from each individual response, add the squared values, and take the square root of the sum. The SRSS technique does not result from any explicit analytical approach but rather is a method to approximate the dynamic coupling of random transients when the time phasing is unknown. Historically, SRSS coupling of dynamic transients has been used in certain conventional building designs and for the design of piping systems in the petrochemical and aerospace industry. SRSS is also used in earthquake engineering, and limited use has been accepted by the NRC in seismic design.

An example of a design equation using the SRSS technique for the combination of SSE and LOCA loads is provided below (N represents normal loads):

$$N + \sqrt{(SSE)^2 + (LOCA)^2} \leq \text{Level D Service Limit}$$

Advantages/Disadvantages

The applicability of the SRSS method as a design tool for use in the design of safety-related structures, systems, and components centers on two issues. The first is the extent to which margins of safety inherent in the ABS method could be reduced considering all other potential concerns such as the possibility of undetected flaws in piping. The second is the relative conservatism or nonconservatism inherent in the SRSS technique to predict the random combination of two dynamic responses. The first of these considerations is largely a matter of regulatory philosophy, i.e., should conservatism be incorporated in the response combination methodology to compensate for other uncertainties in the as-constructed system. The Working Group believed that the method of response combination should be selected on its merits and divorced from what are essentially extraneous factors. In addition, artificial conservatism in the response combination can lead to arbitrary and uncertain margins in the design that may or may not be appropriate for the intended purpose. In any event, the regulatory approach of defense in depth is designed to encompass errors or operational problems such as pipe cracks without the need for arbitrary conservatism in the analysis tools. To address the second consideration, the group surveyed the studies conducted by General Electric and Westinghouse to attempt to quantify the adequacy of the SRSS technique. The General Electric studies were an effort to quantify the probability of exceeding the SRSS value. In its evaluation, GE selected 291 dynamic transients from various events (a non-uniform population) such as LOCA, safety relief valve (SRV) discharge, and earthquake (OBE and SSE). Each of these transients contained random amplitudes and phases, relatively short time durations, and were rapidly varying. A sensitivity study was conducted by GE to investigate the effects on exceedance probabilities for variations in the amplitude ratios, frequency content, and transient durations. This study concluded that the exceedance probabilities are relatively insensitive to changes in these parameters. The transients selected by GE were elastically calculated responses of different components and structures. The conservative assumption of using elastic superposition of the time-phased response transients was also applied in this study. GE reported that the SRSS value was not exceeded for approximately 86 percent of the combinations investigated, which include all of the transient responses identified above. GE also reported that 95.5 percent of the combinations exceeding the SRSS value did so by less than 10 percent of the SRSS value.

Westighouse also conducted studies to quantify a basis for justifying SRSS. Westinghouse looked at only LOCA and SSE response transients for the primary coolant system of its PWRs. Westinghouse used elastically calculated responses that were linearly superimposed by random time phasing. This study also included the normalization of the peak response amplitudes to represent the largest difference between the ABS and SRSS approach. The transients selected by Westinghouse represented three earthquakes (nine transients) and several LOCAs (21 transients total). Westinghouse reported that 75 percent of the combinations investigated were less than the SRSS value.

Both the GE and Westinghouse evaluations of the SRSS method predict a finite probability of exceeding the SRSS value and therefore raise the question of the potential consequences of exceeding the SRSS value. General Electric funded research to investigate and quantify the differences between the response of a structure to dynamic loads and its response to static loads. The study considered the response of a single degree of freedom model and varied parameters such as fundamental frequency (5 and 16 Hz) and ductility ratios (2 to 3). A multi-degree of freedom beam model was also used to demonstrate applicability of this study to multi degrees of freedom systems. Typical earthquakes and SRV transients were used along with some other loading cases such as a triangular load pulse. This study concluded that, even for a worst case model (ductility ratio equal to two), a dynamic reserve margin

(DRM) existed that was 50 percent greater than the static reserve margin (SRM). General Electric further concluded that because the allowable stress limits in the ASME Code were developed based on static load considerations, the dynamic reserve margin would more than compensate for excursions above the SRSS value and the Code safety margin would not be reduced. The discussions in the GE (Reference 1) and Westinghouse (Reference 5) reports could be summarized as follows:

(1) SRSS has been accepted by the NRC staff in some applications such as summing certain seismic modal responses (see Regulatory Guide 1.92).

(2) SRSS accounts for the rapidly varying changes in frequency content and amplitudes of these dynamic events in a more realistic manner than the ABS method.

(3) Use of SRSS in lieu of ABS offsets some of the conservatisms inherent in doing an elastic analysis for structures that would be predicted to behave inelastically. The conservatisms exist due to the energy absorption characteristics of a structure in the inelastic range that are not currently considered in the superposition of elastic responses.

(4) The probabilistic studies conducted by Westinghouse and General Electric both predict a finite (i.e. small but not zero) probability of exceeding the SRSS value.

(5) The probabilistic exceedance predictions contained conservative assumptions but were based on a relatively small data base and, for the General Electric study, on a non-uniform population.

(6) If the probability of the combined dynamic events is low, then the conditional probability that the actual peak combined response will exceed the SRSS calculated peak response is also low and of less significance than for higher probability events.

(7) The difference between the combination of two dynamic responses by ABS and SRSS becomes small as one response peak becomes large with respect to the other response peak. The actual percentage difference between design values (with ABS and SRSS) will be something less than the difference between the dynamic responses because the normal operating loads in the design equations are always added absolutely.

(C) Decoupling of LOCA and SSE Loads

Although the charter of the NRC staff Working Group did not specifically include the study of the interdependency of postulated accidents and natural phenomena, the group's discussions with certain of the industry groups included consideration of this subject. As discussed earlier in this report, the consideration of combined accident and earthquake loads derives from GDC 2. It has been the staff interpretation that systems important to safety be designed against LOCA plus SSE loads although there has not been a clear articulation of the intent of this requirement, i.e., is there intended to be an explicit physical dependence between the LOCA and SSE? If the separate events are considered to be statistically independent, then it can be shown that the probability of a LOCA event and SSE event occurring in the same general time frame is extremely low and that the probability of the peak LOCA response and peak SSE response combining is exceedingly low and of a magnitude which would not be indicative of a design basis event. On the other hand, calculations of the potential coupling of the events, i.e., SSE causing a LOCA, also yield low probabilities although the numbers are subject to considerable controversy (Reference 6).

At the time of NUREG-0484 report, Westinghouse had made the most visible progress in developing a technical basis for decoupling LOCA and SSE loads, including sub-mittal of a topical report (Reference 6) to the NRC. B&W and CE indicated that low-level efforts were being maintained and that there was great interest in whether the NRC would be receptive to further work in this area.

The Westinghouse studies investigated the likelihood of a large seismic event causing a LOCA and concluded that an SSE could not reasonably be the initiator of such an event. One analysis postulated an undetected flaw in the pipe wall and showed that seismic loadings would not produce significant fatigue crack growth. A further analysis postulated a large through wall flaw (which would ordinarily be detectable) and showed that extension of the flaw due to instability would not occur.

It was the consensus of the Working Group that the studies performed to date (September 1978), such as by Westinghouse, are promising, although further work

is necessary to confirm an adequate technical basis for decoupling. The group
believed that pursuit of this additional work should be encouraged by the NRC.

(D) 1978 Working Group Recommendations - NUREG-0484

Based on studies of proposed response combination methodologies, the 1978 NRC
Working Group concluded that a limited application of the SRSS method could be
supported at that time,[3] and, therefore, they developed recommendations for
combining dynamic responses in the analyses of structures, systems, and
components important to safety. These were published in September 1978 in
NUREG-0484 (Reference 8).

NRC in September 1978 adopted the Working Group's recommendations which stated:

"For the combination of the dynamic responses within the Reactor Coolant
Pressure Boundary (RCPB) and its supports, which result from the coincidence
of an SSE and LOCA, the Square Root of the Sums of the Squares (SRSS)
technique is acceptable contingent upon performance of a linear elastic
dynamic analysis to meet the appropriate ASME Code, Section III, Service
Limits."

---

[3]This was in addition to the uses of SRSS that the NRC had already approved
prior to acceptance of NUREG-0484 recommendations.

3. SSE AND LOCA RESPONSES - RCPB AND BALANCE OF PLANT

The study described in NUREG-0484, dated September 1978, was limited to the primary reactor coolant system because the Working Group focused only on the available information regarding the design methods and responses of various primary system components and dynamic loads. Since the publication of NUREG-0484, the staff has considered the design and responses of ASME Section III Class 2 or 3 systems, and compared the effects of applying SRSS and Absolute Sum (ABS) methodologies to BWR applications for Operating Licenses. The staff has obtained additional information from the work done at General Electric Company[4] (Reference 9), the Mark II Owner's Group and their consultants Dr. R. P. Kennedy and Dr. Nathan Newmark[4] (References 10 and 11). The staff also has had the benefit of generic dynamic response combinations studies at Brookhaven National Laboratories[5] (Reference 12).

Evaluation of this information indicates that ABS method may result in a design overly influenced by unlikely peak combinations rather than a design which considers modes of operation. Design by using ABS method may make these systems less reliable for normal operation, and is particularly a concern for piping systems which are subjected to large thermal expansions and relative motions that are best accommodated by flexibility. SRSS represents a method of

---

[4]NEDE 24010-P, NEDE 24010-1, NEDE 24010-2, NEDE 24010-3 (including tapes of various transients supplied by MK. II Owner's Group to NRC & BNL).

[5]NUREG/CR-1330.

designing a less rigid system than ABS and thus permits a degree of design optimization when thermal stresses are present and excessive rigidity is not desirable.

The evaluation of the 1978 Working Group recommendations (Section 2(D)) and the technical bases (see paragraphs (A) thru (E)) has led the staff to the conclusion that the SRSS methodology for combining responses[6] due to LOCA[7] and SSE can be extended to any ASME Class 1, 2, and 3 affected system, component, or support.

In applying SRSS methodology, the responses should be calculated on a linear elastic basis and the time-phase relationship among functions to be combined should be random. Functions which are not independent such as those that result from the same dynamic initiating event should not be combined by SRSS methodology. In those cases for which the independence of the function cannot be established (e.g., Annulus pressurization and jet impingement or reaction loading which result from a postulated LOCA), absolute summation of responses or the use of time-phased responses is required.

The technical bases for this extension of SRSS methodology to include any ASME Class 1, 2, and 3 system, component, or support is summarized below in paragraphs (A) through (E), and is also supported by the work performed by

---

[6]The individual responses have rapidly varying amplitude and a limited number of short duration peak responses with the proviso that the frequency content of the associated forcing functions provide significant excitation of the natural frequencies of the component.

[7]Pipe breaks in RCPB that result in the maximum response in any component or component support.

Brookhaven National Laboratories for NRC reported in NUREG/CR-1330. Due to the lack of of comparable ductility for concrete structures, and procedural differences between Divisions 1 and 2 of the ASME Section III Code in considering loads and load combinations, the application of SRSS methodology to structures has not been accepted at this time.

(A) Low Probability of Occurrence.

The individual probabilities of occurrence of the SSE and LOCA are small. If these two events are statistically independent, the probability of simultaneous occurrence of these events would be very low. Alternatively, given the occurrence of an SSE, there is some finite probability that a LOCA could result. Although difficult to quantify at this time, G.E. and Westinghouse studies have indicated that this conditional probability is also low. Within this low probability frame of reference, the possible excursions of the combined dynamic response over the SRSS value are of less significance than would be the case for higher probability events. (In this regard the use of low versus high probability is intended to distinguish between design basis accidents as contrasted to types of events that are anticipated to occur at least once in the plant lifetime.)

(B) Engineering Judgment Indicates that SRSS Represents a More
    Realistic Combination of Peak Dynamic Responses for Rapidly
    Varying Frequencies and Amplitudes, Short Time Duration of
    Peaks and Random Time Phasing

The nature of the dynamic response time histories, e.g., rapidly varying
frequencies and amplitudes, short time durations of peaks, and random time
phasing would indicate that an ABS combination is very unlikely, even though
this combination can be attained on a theoretical basis. In addition, the
techniques used to calculate SRSS nonexceedance probabilities conservatively
assume that peak responses superimpose linearly when the structure behaves
inelastically. This assumption does not account for the expected coupling
behavior for an inelastic structure that exhibits higher energy absorption
capabilities than would be predicted by the elastically calculated response.
Good engineering judgment is therefore supportive of the SRSS method as being
a more realistic combination of peak dynamic responses.

(C) Techniques Used to Calculate SRSS Nonexceedance Probabilities
    Contain Conservative Assumptions

Both Westinghouse and General Electric utilized Probability Distribution
Functions (PDF) and Cumulative Distribution Functions (CDF) in their studies
to quantify SRSS nonexceedance probabilities. Examples of conservatisms
utilized in these techniques include normalization to 1.0 of each individual
response that yields the maximum difference between ABS and SRSS summation of
peak responses and random statistical superposition of the peak responses so

- 17 -

that the maximum sum of the absolute values of the peaks can be attained. In addition, CDF parametric sensitivity studies were performed that demonstrated that the nonexceedance probabilities of SRSS were relatively insensitive to the parameters that could affect the CDF distribution.

(D) Conservatisms Inherent in Linear Elastic Dynamic Analysis

ASME Code, Section III, stress limits are equivalent static limits and therefore do not account for the additional energy absorption capability of components under short duration (dynamic) energy limited forcing functions. This fact is generally recognized by the engineering community and is not unique to this study. Work undertaken to quantify this additional energy absorption capability has shown that for components exhibiting even moderately ductile behavior, the Dynamic Reserve Margin (DRM) is generally much larger than the Static Reserve Margin (SRM). Because the use of SRSS is contingent upon performance of a linear elastic dynamic analysis, the additional dynamic design margin (i.e., $\frac{DRM}{SRM}$) appears to compensate for peak combined responses which could probabilistically exceed the SRSS value.

(E) Independence of Dynamic Response Combination Techniques from Other Factors Affecting Design Margins

The design margins inherent in the fluid systems of nuclear power plants are not only dependent upon reasonably accurate load definition and structural analyses, but encompass material integrity programs, adequate fabrication and examination techniques, effectiveness of inservice inspection and many other

factors. Periodically, operating experience also uncovers problems such as
snubber malfunctions, steam generator tube wastage, and flaws or cracks in
safety-related components that have some cumulative effect on overall design
margin. It has been contended that conservative assumptions should be made
regarding response combinations to provide additional design margin against
such contingencies. The staff believes that this is not a valid consideration, principally because the amount of margin achieved through the selection
of a particular load/response combination methodology is quite variable and
may or may not be appropriate. Therefore, the staff recommendation is based
on an assessment of the technical merits of how dynamic responses should be
combined, rather than the need or desirability of providing design margin,
against those contingencies.

4.   CRITERIA FOR COMBINATION OF DYNAMIC RESPONSES OTHER THAN THOSE OF SSE
     AND LOCA

General Electric Company, the Mark II Owner's Group and their consultants
presented information from their studies which concluded that there are several
bases (reliability, optimum design, dynamic margin and statistical basis) for
accepting the SRSS combination of peak dynamic responses. The Brookhaven
National Laboratories (BNL) under Technical Assistance Contract from NRC
studied the statistical basis and confirmed that the rules of SRSS methodology
are valid provided that the loads are conservatively defined and the
characteristics of response functions are carefully examined. NUREG/CR 1330
contains a series of demonstration analyses using simulated time functions and

actual Mark II response combination cases. All responses[8] received from General Electric and the Mark II Owner's Group were evaluated.

These Mark II combinations evaluated were the dynamic responses to OBE, SSE, combined with SRV and CHUG load, for the RHR wetwell piping systems, the feedwater piping, the main steam line piping, the reactor pressure vessel skirt, the shroud support and the guide tube.

The results of these evaluations formed the bases for accepting SRSS methodology and are as follows:

- The Non Exceedence Probability (NEP) for the SRSS combination of two responses generally has a value greater than 0.50. Specifically, 30 percent of such combinations were greater than 0.90, 80 percent greater than 0.60, 92 percent greater than 0.50, and 8 percent less than 0.50.

- For the cases where the NEP using SRSS was below 0.50, the difference between SRSS and ABS was observed to be small, and the peak magnitude of one of the responses was observed to be much larger than the peak magnitude of the other response. For such cases, the use of a multiplication factor of 1.2, applied to the peak magnitude of the responses will result in an adjusted combination value which will be in an acceptable range.

---

[8]All responses with the exception of those pertaining to the containment structure.

Based on the staff study of the Newmark-Kennedy (N-K) Criteria (Reference 10), its supporting document (Reference 11), and the Brookhaven National Laboratories work reported in NUREG/CR 1330, the staff has arrived at the following position which is applicable to all mechanical systems, components, and supports for combining responses to dynamic loads other than LOCA and SSE.

The SRSS method may be used when Conditions A and B are both satisfied.

Condition A

    (i)   The dynamic response time function is rapidly varying;

    (ii) Duration of the strong motion portion of the function is short;

  (iii) Function consists of a few distinct high peaks which are random with respect to time;

    (iv) Response is calculated on a linear elastic basis; and

    (v)   Time-phase relationship among functions to be combined is random.

Condition B

For loads which meet Condition A, the SRSS method may be used provided a non-exceedance probability (NEP) of 84 percent or higher is achieved for the combined response. An acceptable method of attaining that goal is meeting all the following requirements:

(i) Define loads at approximately the 84 percentile or 1.15 times the median, whichever is greater.

(ii) The SRSS value of the response combination has an NEP $\geq$ 50 percent selected from a Cumulative Distribution Function (CDF) curve constructed on the assumption that individual response amplitudes are known and only random time phasing defined by its probability density function exists. The CDF curve may be developed using the procedures of Appendix N of Section III of the ASME Code, or alternatively methods developed by BNL (Reference 12), or Westinghouse in WCAP-9279 (Reference 5) using absolute (unsigned) values of response amplitude may be used. The method selected shall be justified in the submission for the application being analyzed.

(iii) 1.2 times the SRSS value of the response combination has an NEP $\geq$ 85 percent from the CDF curve constructed as in (ii) above assuming only random time phasing.

The confirmatory work in NUREG/CR 1330 (Reference 12) also has demonstrated that the SRSS use of methodology accepted for the combination of SSE and LOCA responses by both Sections 2 and 3 of this report will provide a non-exceedance probability of 84 percent or higher. For the combination of LOCA and SSE responses, it is not necessary to apply the procedure described in this section

(Section 4) since we have determined that Condition A has been satisfied and that the 84 percent NEP level has been satisfied.

Should a case not meet the response combination criteria stated in Conditions A and B above, responses shall be combined by absolute sum, or alternatively other methods may be used provided justification is furnished to the staff to ensure that the 84 percent level for NEP of the response combination has been met.

5.   CONCLUSIONS

The staff considers the use of SRSS appropriate for:

(i)   Combination of SSE and LOCA dynamic responses for all ASME Class 1, 2, or 3 systems, components, or supports. For dynamic responses resulting from the same initiating event, when time-phase relationship between the responses cannot be established, the absolute summation of these dynamic responses should be used.

(ii)   Combining responses of dynamic loads other than LOCA and SSE provided a non-exceedance probability (NEP) of 84 percent or higher is achieved for the combined SRSS response. An acceptable method for achieving that goal is outlined in Section 4, Condition A and Condition B, paragraphs (i), (ii) and (iii).

6. BROOKHAVEN NATIONAL LABORATORY GENERIC STUDIES ON RESPONSE COMBINATION METHODOLOGY STATUS

Brookhaven National Laboratory under contract from NRC is continuing its work on response combination methodology beyond that documented in NUREG/CR 1330. The second phase (Phase 2B of TAP B-6)[9] of the BNL work will investigate the validity of several proposed "loading level" criteria reported in NUREG/CR 1330. A "loading level" criterion that permits a determination of the applicability of SRSS methodology to be made by simple inspection of the loading input is much to be preferred over a "response level" criteria contained in the position outlined in Section 5.

The broader subject of which dynamic events and natural phenomena require combination, how responses in components, equipment, and structures should be combined and the specification of suitable design margin for such combinations is under active review within the scope of Task Action Plan B-6, "Loads, Load Combinations and Stress Limits."

---

[9]Task Action Plan B-6, Section 2B.

DEFINITIONS

1. Cumulative Distribution Function (CDF): The integral of the Probability Density Function over a range from 0 to X, i.e., CDF $(x) = \int_0^X P(x)dx$

2. Dynamic Design Margin (DDM): The ratio of the Dynamic Reserve Margin (DRM) to the Static Reserve Margin (SRM), i.e., $DDM = \frac{DRM}{SRM}$

3. Dynamic Reserve Margin (DRM): For dynamic loads DRM is the ratio of the peak dynamic load to cause failure, calculated by dynamic elastic-plastic analysis, to the peak dynamic load resulting in code stress limits calculated by linear elastic analysis.

4. Juctility Ratio (μ): Total allowable deformation of a structural system divided by the effective yield deformation.

5. Probability Density Function (PDF): In the general case PDF is P(X) as a function of X where X is defined as a time of event occurrence, response magnitude, etc.

6. Static Reserve Margin (SRM): For static loads SRM is the ratio of the static load to cause fuilure, calculated by elastic-plastic analysis, to the static load resulting in Code stress limits calculated by linear elastic analysis.

REFERENCES

1. NEDE-24010-P, Technical Bases for the Use of the Square Root of the Sum of Squares (SRSS) Method for Combining Dynamic Loads for Mark II Plants, July 1977.

2. Technical Bases for the SRSS Method of Combining Dynamic Responses, General Electric Co., April 1978.

3. EDAC 134-240.7R, Combining Multiple Dynamic Responses by the Square Root of the Sum of the Squares Method, May 1978.

4. Probability Evaluation for Dynamic Response Combinations, Nuclear Services Corp., 1700 Dell Avenue, Campbell, CA 95008.

5. WCAP-9279, Combination of Safe Shutdown Earthquake and Loss-of-Coolant Accident Responses for Faulted Condition Evaluation of Nuclear Power Plants, March 1978.

6. WCAP-9283, Integrity of the Primary Piping Systems of Westinghouse Nuclear Power Plants During Postulated Seismic Events, March 1978.

7. SDAR-78-02, Inherent Design Margins in Structures Subjected to Pulse Type Loads, May 1978.

8. NUREG-0484, NRC Working Group Report, "Methodology for Combining Dynamic Responses."*

9. NEDE-24010-1, Supplement 1, SRSS Application Criteria as Applied to Mark II Load Combination Cases, October 1978.

10. NEDE-24010-2, Supplement 2, Bases for Criteria for Combination of Earthquake and Other Transient Responses by the Square-Root-Sum-of-the-Squares Method, December 1978.

11. NEDE-24010-3, Supplement 3, Study to Demonstrate the SRSS Combined Response has Greater than 84 Percent Non-Exceedence Probability When the Newmark-Kennedy Acceptance Criteria are Satisfied, August 20, 1979.

12. NUREG/CR-1330, Brookhaven National Laboratory Review of Methods and Criteria for Dynamic Combinations in Piping Systems, Final Report, April 1980.*

---

*Available for purchase from the NRC/GPO Sales Program, U.S. Nuclear Regulatory Commission, Washington, D.C. 20555, and the National Technical Information Service, Springfield, Virginia 22161.

NRC FORM 335 (7-77) U.S. NUCLEAR REGULATORY COMMISSION BIBLIOGRAPHIC DATA SHEET	1. REPORT NUMBER (Assigned by DDC) NUREG 0484, Revision 1

4. TITLE AND SUBTITLE (Add Volume No., if appropriate) Methodology for Combining Dynamic Responses	2. (Leave blank)
	3. RECIPIENT'S ACCESSION NO.

7. AUTHOR(S)	5. DATE REPORT COMPLETED
	MONTH: April YEAR: 1980

9. PERFORMING ORGANIZATION NAME AND MAILING ADDRESS (Include Zip Code) (Unresolved Safety Issues Program) U. S. Nuclear Regulatory Commission Office of Nuclear Reactor Regulation Washington, D.C. 20555	DATE REPORT ISSUED
	MONTH: May YEAR: 1980
	6. (Leave blank)
	8. (Leave blank)

12. SPONSORING ORGANIZATION NAME AND MAILING ADDRESS (Include Zip Code) Division of Safety Technology Office of Nuclear Reactor Regulation U. S. Nuclear Regulatory Commission Washington, D.C. 20555	10. PROJECT/TASK/WORK UNIT NO.
	11. CONTRACT NO.

13. TYPE OF REPORT Technical Report	PERIOD COVERED (Inclusive dates)

15. SUPPLEMENTARY NOTES	14. (Leave blank)

16. ABSTRACT (200 words or less)

Procedures in accordance with Appendix A of 10CFR50, GDC 2, call for an appropriate combination of the effects of the accident loads and loads caused by natural phenomena (such as earthquakes) to be reflected in the design bases of safety equipment. This requirement of interaction of loads has been implemented in various ways both within the NRC and the Nuclear Industry. An NRR Working Group constituted to examine load combination methodologies developed recommendations which were published in September 1978 as NUREG-0484. NRC adopted the Working Group's recommendations which stated: "For the combination of the dynamic responses within the Reactor Coolant Pressure Boundary (RCPB) and its supports which result from the coincidence of an SSE and LOCA, the Square Root of the Sums of the Squares (SRSS) technique is acceptable contingent upon performance of a linear elastic dynamic analysis to meet the appropriate ASME Code, Section III, Service Limit." Revision 1 of NUREG-0484 extends the conclusions of the original NUREG-0484 on the use of SRSS methodology for the combination of SSE and LOCA responses beyond RCPB to any other ASME Section III, Class 1, 2, or 3 affected system, component or support, and provides criteria for the combination of dynamic responses other than SSE and LOCA.

17. KEY WORDS AND DOCUMENT ANALYSIS	17a. DESCRIPTORS

17b. IDENTIFIERS/OPEN-ENDED TERMS

18. AVAILABILITY STATEMENT Unlimited Availability	19. SECURITY CLASS (This report) Unclassified	21. NO. OF PAGES
	20. SECURITY CLASS (This page) Unclassified	22. PRICE S